WARSHIP PICTORIAL 39

GERMAN BATTLESHIPS 1939-1945

by Steve Wiper

BISMARCK fitting out, pierside, at the Blohm & Voss Shipyard, Hamburg, Germany, July-August 1940.

CLASSIC WARSHIPS PUBLISHING

P. O. Box 57591 • Tucson, AZ 85732 • USA

Web Site: www.classicwarships.com • Ph/Fx (520)748-2992

ISBN 978-0-9857149-1-8

Printed by Arizona Lithographers, Tucson, Arizona

GERMAN BATTLESHIPS 1939-1945

This book is designed to provide the reader with a chronological photo history of the four German battleships throughout the Second World War. They would be the battleships of the SCHARNHORST class, including GNEISENAU and of the BISMARCK class, including TIRPITZ. All of the battleships were named after famous military leaders from Germany's past, and so, because the traditional calling of a warship as "she" was too feminine for ships named after famous men, they were called "he" by their officers and crew.

As of 1 September 1939, with Germany's invasion of Poland, the Second World War in Europe had begun. At that date, Germany had the battleships of the SCHARNHORST class that were essentially battle-ready. SCHARNHORST was in the Baltic Sea training just after the completion of the *Atlantik* bow modification, so in reality, GNEISENAU was truly the only operational battleship in the German Navy at that time.

Germany's opponents at the beginning of the war that had a navy of any significance were Great Britain and France. The British Royal Navy(RN) had 12 battleships and 3 battlecruisers and the French Navy had 6 battleships and 2 battlecruisers. The German Navy was hopelessly outnumbered and therefore had no plans to face either opponent in the traditional sense. The plan for the German Navy was to perform commerce raiding to try to starve their enemies to submission. At that time, two battleships of the BISMARCK class were building, as well as two of the larger "H" class.

Once SCHARNHORST had completed trails and training, he and GNEISENAU made a foray into the North Atlantic Ocean, off the coast of Norway during October 1939, with no results. Later, in November, both battleships sortied again to the waters south of Iceland, sinking the RN auxiliary cruiser RAWALPINDI. December was spent by the two battleships in port making modifications due to damage at sea.

January 1940, both battleships were iced-in at Kiel due to a very severe winter, with no action. Come February, icebreakers freed the two battleships so that they were able to depart for Operation Nordmark, to intercept British convoys between England and Norway, with no success. In April 1940, SCHARNHORST and GNEISENAU set out into the Atlantic Ocean again in Operation Weserübung, as the naval cover force for the invasion of Norway. They briefly encountered the RN battlecruiser RENOWN, with GNEISENAU receiving slight damage. Later in June, both battleships ventured into the North Atlantic, Operation Juno, sinking the RN aircraft carrier GLORIOUS and two destroyers, but SCHARNHORST was damaged by a torpedo hit. In late June GNEISENAU was damaged by a torpedo hit. Both German battleships were then out of operation and had returned to Germany for repairs, leaving Germany with no battleships in operational status.

Hitler's newest battleship, BISMARCK, was not commissioned until 24 August 1940, and still had a long work-up period to go through before he could become a viable warship. TIRPITZ, built almost in parallel, would not commission until 25 February 1941, due to up-grades and changes made, that were learned from the construction and work-up of BISMARCK. Both the SCHARNHORST and GNEISENAU were operational again by the end of 1940, both failing to break out into the Atlantic due to extreme weather conditions.

In January 1940, SCHARNHORST and GNEISENAU were able to sortie into the Atlantic Ocean for their longest voyage, Operation Berlin, which would last for two months. This convoy raiding cruise was in the North and Central Atlantic Ocean, sinking 22 merchant vessels and the battleships steamed just over 16,000 nm. Both battleships put into Brest, France in late March 1941, where they stayed until February 1942.

BISMARCK, worked up and trained in the Baltic Sea, was ready for operational duty in May 1941, sortied with heavy cruiser PRINZ EUGEN into the Atlantic for his first mission. They were intercepted by HMS HOOD and PRINCE OF WALES. HOOD was sunk and the biggest sea chase of WWII ensued. BISMARCK was wounded by aircraft torpedoes and caught in the Central Atlantic by and overwhelming force and sunk. Soon after, TIRPITZ was ready for operational duty, but retained in the Baltic due to the fear of losing him in the Atlantic, which had by that time become too risky for German battleship operations. In November 1941, the two "H" class battleships under construction were canceled.

January 1942 saw TIRPITZ transferred to Norway for operations against Russian bound convoys in the North Atlantic and Arctic Seas. By this time, the United States interred the war, committing at least 7 seven battleships, combined with at least 10 from Great Britain, the German battleships were under a constant threat from overwhelming forces and would thereafter spend the majority of their time in harbor. This is where the threat of aerial attack was dramatically increased. February saw the Channel Dash, or Operation Cerberus, the escape of SCHARNHORST and GNEISENAU from the constant threat of aerial bombardment at Brest, France, back to Germany. Both battleships hit mines and required major repairs. GNEISENAU was bombed two weeks later and heavily damaged. SCHARNHORST would need repairs at Kiel and Gotenhafen for ten months from the mine damage. For most of 1942, TIRPITZ was the only operational German battleship. Due to this, the British nick-named him the "Lonely Queen of the North." TIRPITZ was then under many aerial attacks while hiding in the Norwegian fjords. In March, TIRPITZ sortied to attack a convoy, but failed to find it and returned to Norway. In July, TIRPITZ sortied to attack another convoy, but the convoy scattered and TIRPITZ returned to Norway. Every month the battleship was under air attack. GNEISENAU was to be highly modified due to the massive repairs needed, so he was to remain at Gotenhafen, without a bow, damaged by the previous bombing.

January 1943, SCHARNHORST had completed repairs and post-repair trials and was operational again. SCHARNHORST joined TIRPITZ in Norway in March, but the two battleships would only perform training through August 1943. In September TIRPITZ and SCHARNHORST bombarded in support of the invasion of the island of Spitsbergen, the Allied weather station vital to convoys to Russia. Later that month, TIRPITZ suffered major damage from mines placed by RN midget submarines, and was under repair until March 1944. SCHARNHORST sortied against an Allied convoy to Russia in December, only to be sunk in the Arctic Sea by an overwhelming British force of a battleship, cruisers and destroyers.

March 1944 saw TIRPITZ operational again and the only German battleship to be so. Because of this, he was under constant aerial attack, sometimes several times a month. In April, TIRPITZ was seriously damaged, necessitating repairs for months. In September, TIRPITZ was hit by a 5,000 lb. bomb in the bow and again severely damaged. It was determined that the battleship could not be fully repaired and that the ship should be anchored in very shallow water, where it could not be sunk. In November 1944, TIRPITZ was bombed by the RAF with 12,000 lb. bombs and sunk. During the remaining months of the Second World War, the last German battleship was GNEISENAU, de-commissioned, modification work halted, he was eventually scuttled at Gotenhafen, late March 1945.

It was amazing what Germany was able to do with so few battleships as compared to their enemies. It was also incredible that Germany was able to operate and retain their use for as long as they did. At any given time, they were outnumbered at least 6 to 1, and at times, 11 to 1, by battleships of their adversaries, a threat that would be the demise of BISMARCK and SCHARNHORST. Being under near constant aerial bombardment from the British Royal Air Force (RAF), was a threat that they had to contend with, and were not always successful in defending against. This threat would be the demise of GNEISENAU and TIRPITZ. In the end, Germany's battleship campaign during the Second World War was a losing venture with the numbers they had in these warships. To have operated with any effectiveness, Germany would have required at least eight battleships.

SCHARNHORST, at sea, possibly Kiel Bay, late August 1939, on the eve of war in Europe. The battleship just completed a refit, with the addition of the *Atlantik* bow, anchor seats, enlarged hangar, raked funnel cap and the mainmast relocated to the after superstructure. He would work-up and perform battle training in the Baltic Sea into November 1939.

GNEISENAU at anchor just before the beginning of the war on 1 September 1939. He had many of the same modifications as SCHARNHORST, but several months prior. The mainmast was not relocated, the aircraft hangar midships was a different configuration and the boat/aircraft cranes were built to a slightly different shape to that aboard his brother.

BISMARCK, pictured here, sometime during September 1939, with his construction well along. The *Atlantik* bow was fitted, but awaiting the forward bullnose. BISMARCK was launched with the older straight stem, a design with origins going back to German warships of the "Great War." With a new doctrine of commerce raiding on the high seas, the *Atlantik* bow design allowed the warship to slice through the sea and ship less water over the foredeck. Note that the hull armor belt was being installed at the date of this photograph, and that turrets Anton (1) and Bruno (2) were still under construction.

GNEISENAU, amidships, sometime September 1939. Note the difference in the crane compared to that aboard SCHARNHORST, pictured right.

SCHARNHORST, with the crew manning the rails, welcoming U-47 home, 23 October 1939, after that submarine had sunk the British battleship HMS ROYAL OAK at Scapa Flow. The medium gray paint on the hull was the standard for the Kriegsmarine, Dunkelgrau 51, a German Navy designation. This paint color was also known as RAL 7000, a German government paint number designation. The superstructure and everything above the height of the main deck was painted a light gray known as Hellgrau 50, or RAL 7001. This pattern was the standard for the German Navy(s) since 1896.

GNEISENAU, at anchor during the very harsh winter of 1939-40, Kiel Bay, Germany. The aircraft catapult is still fitted to the after turret, known as Caesar in the Kriegsmarine, which would date this photo about December 1939, or January 1940. Note the additional catapult just forward of the 10.5m after rangefinder. The feature that distinguished GNEISENAU the most from SCHARNHORST was the fitting of the mainmast to the backside of the funnel. Originally, both battleships were built with the mainmast in this location, but SCHARNHORST would later have it relocated to the after superstructure.

Another photograph of GNEISENAU at anchor in Kiel Bay during the winter of 1939-40. Note the dark color in this photo on the top of his forward turrets. At this time GNEISENAU had the recognition color of Deckfarbe Kaiserrot (Deck Color King's Red), or RAL 3010 paint atop the forward turrets in preparation for Operation Nordmark. That would date this photo about January, or early February 1940.

SCHARNHORST tied up at the Fliegerdeich Pier in the Hipper Hafen, Kiel, Germany, January 1940. The 10.5m main battery rangefinder atop the bridge structure was fitted with the then new FuMO-22 fire control radar antenna the previous month. A structure was fitted atop the original rangefinder to adopt this radar system.

GNEISENAU, with his brother SCHARNHORST, and the heavy cruiser BLÜCHER just beyond, Kiel Bay about February 1940.

SCHARNHORST, 27 February 1940, just returned from Operation Nordmark, led by his brother GNEISENAU as Flagship, was an attempt to hunt British convoys, but no ships were located, so the German squadron returned to port.

This photo is of the SCHARNHORST just after returning from Operation Nordmark, late February 1940. The deck of the warship in the foreground is that of GNEISENAU. Note the waterway along the deck edge. This was fitted to shed water off the cambered main deck.

The image on the right is of the GNEISENAU in March 1940. Note the difference of his bridge structure from that of SCHARNHORST. At this time, both battleships were still painted in the standard 50(superstructure), 51(hull).

SCHARNHORST, with his brother GNEISENAU ahead, which was the Flagship for Operation Weserübung, along with the heavy cruiser ADMIRAL HIPPER and 14 destroyers in the invasion of Denmark and Norway, 7-12 April 1940. The two battleships were to provide distant cover for the invasion operation. They had a brief encounter with the RN battle-cruiser HMS RENOWN, with both RENOWN and GNEISENAU damaging each other lightly. At the time of this photograph, the recognition color atop the main and secondary gun turrets was RAL-1003, or Deckfarbe Gelb(Yellow).

This photograph is of the GNEISENAU, possibly after Operation Weserübung, in port. The turrets would have been painted with a recognition color of RAL-3010 at that time. Note the Arado Ar-196 floatplane mounted atop the catapult. At this time, this aircraft was painted on the upper surfaces with RLM-73 and the undersides with RLM-65.

The SCHARNHORST, led by GNEISENAU, about late May 1940 in the Baltic Sea. The two battleships were training their gun and flak(anti-aircraft) mount crews. GNEISENAU was also testing his then newly installed anti-mine degaussing gear. Both battleships were operating out of the Polish port of Gotenhafen, which was the primary German Navy port for the Baltic, as it was out of reach of Allied bombers. Main and secondary turret tops were painted with the RAL-3010 red recognition color at that time and the battleships were painted in the standard 50/51 pattern.

SCHARNHORST, leading GNEISENAU during Operation Juno in the North Atlantic, their most famous battle against the British Royal Navy. They encountered the RN aircraft carrier HMS GLORIOUS and two escorting destroyers. Needless to say it was a one-sided battle. GLORIOUS and both destroyers were sunk, but SCHARNHORST was hit by one torpedo, causing significant damage. The pair of German battleships put into the Norwegian port of Trondheim for emergency repairs. Turret tops were painted a dark gray, RAL-7016 during this operation.

BISMARCK in a floating drydock, 23 June 1940, during his final stages of the fitting-out process at the Blohm & Voss Shipyard in Hamburg, Germany. Note the graceful flare to the "Atlantik" bow and the bulbous forefoot of the lower bow. The "nub" at the forefoot was the cover for the extension used for minesweeping. The massive girth of the BISMARCK class hull can be seen here.

A view of the after superstructure of the SCHARNHORST after that battleship returned to Kiel, Germany for extensive repairs to his hull, both center and starboard engines, as well as damages to turret Caesar. The date of this photograph was during the summer of 1940. The rangefinder visible at the top of this photograph was a 10.5m optical device. The battleship remained painted in the 50/51 pattern at this time and the turret tops were painted dark gray RAL-7016 while under repair at the shipyard.

GNEISENAU, while anchored at Trondheim, Norway, due to submarine torpedo damage to his bow on 20 June 1940, by HMS CLYDE. These images were taken in late June, early July. This image is from July with the camouflage pattern during that time.

This photograph was taken in late June with the initial camouflage pattern on GNEISENAU. Note the work to his bow, near the waterline. The hole from the torpedo hit was very large and clear through the other side of the battleship.

The repair ship HUASCARAN was tied up alongside of GNEISENAU in this late June 1940 photograph. The paint colors used for this pattern appear to be RAL-7024 for the lighter and RAL-7016 for the darker of the two shades visible here.

GNEISENAU en route back to Germany, 25 July 1940. The torpedo boat LUCHS was sunk by the RN submarine HMS THAMES, which missed GNEISENAU. The camouflage on the hull was modified and that on the superstructure was painted back to 50, while the recognition color atop the turrets was red RAL-3010.

BISMARCK in the final stages of the fitting out process at the Blohm & Voss Shipyard, Hamburg. The time frame of this photograph was August 1940. He was painted at this time in the standard pattern of 50/51, but the funnel cap was painted aluminum. Note that the battleship is riding high in the water as most of the boot topping is exposed. Also note the similarity of BISMARCK's Admiral's bridge to that of GNEISENAU's on page 9.

BISMARCK, 15 September 1940, on the Elbe River, departing Hamburg Germany, en route to the Baltic Sea to perform machinery trials. Note the girth of this class of battleship, which at 118 ft., or 36m, was the largest of any battleship afloat at that date. That made the BISMARCK class battleship a very stable firing platform. Also at the time of this photograph, BISMARCK had yet to be fitted with the main fire control rangefinders and some of the anti-aircraft rangefinders.

The two photographs above are of the SCHARNHORST wearing a then new and experimental camouflage pattern using black and white stripes to break-up, or alter the silhouette of a vessel, in other words a "disruptive" camouflage pattern. This was applied about October, or November 1940 and was used for only a short time. The recognition color atop the turrets was RAL-3010, while the ship still wore the overall paint pattern of 50/51.

In this bottom photograph, the battleship TIRPITZ, brother to the BISMARCK, was fitting out at the Kriegsmarine Werft Shipyard in Wilhelmshaven, Germany, December 1940. He was painted in a "concealment" pattern, using colors that matched the structures beyond the battleship, namely brick red and brown, over the standard 50/51 gray pattern.

At about the same time frame as the SCHARNHORST received the first of what was to become known as the Baltic Camouflage, GNEISENAU would too receive such a pattern. What we see in these photographs is typical of the final version of this type of camouflage. GNEISENAU had the bow and stern of the battleship painted with RAL-7016 to give the warship a shorter appearance. False bow and stern waves were also applied to confuse an observer's estimation of speed. At this time, the recognition color atop the turrets was red RAL-3010, also a standard for Baltic operations.

BISMARCK in a very rare photograph only recently discovered, wearing a camouflage pattern of the standard 50/51, and in addition, a large portion of the forward hull and a smaller portion of the stern painted with a slightly darker gray, RAL-7024. This pattern may have only been used for one, or two weeks, about early October 1940, during trials in the Baltic Sea.

Another photograph of BISMARCK, this one from January 1941, where the battleship was returned to the standard 50/51 paint pattern. Note that he had been fitted with the 10.5m rangefinders on the foretop and after superstructure, which had the FuMO-23 fire control radar antenna installed on both units. The 7m rangefinder scheduled for installation atop the conning tower had yet to be fitted at the date of this photograph.

GNEISENAU, sometime early in Operation Berlin, 22 January through 22 March 1941. He was painted in the standard 50/51, with the odd addition of a slightly darker stern. This may have been due to poorly applied paint, as the "Baltic" stripes were showing through midships, and what we see here may be the show-through of the previous pattern. At this time he also had the turret tops painted with the recognition color of yellow RAL-1003, a standard for Atlantic operations. The crew was mustered aft to either cover the deck swastika, or the battleship was performing re-fueling operations.

BISMARCK, transiting the Kaiser Wilhelm Canal, en route to the Baltic Sea, 8 March 1941. He was then painted with a version of the "Baltic" camouflage pattern, but with variations to the recognition painting on the turret tops and the main gun barrels.

It would appear that the tops of main gun turrets Bruno(2) and Caesar(3) were painted with dark gray RAL-7016, but turrets Anton(1) and Dora(4) were painted differently. On those turrets, the horizontal tops were painted black, but what color were the slopped sides of the turrets painted? There is a school of thought that they were painted with the standard Baltic recognition color of red RAL-3010. Note the painting of the gun barrels of the B and C turrets where they super-imposed the roof of the A and D turrets. Painted deck swastika were also present fore and aft, with a red background ?

SCHARNHORST near the final days of Operation Berlin, the longest and most successful commerce raiding sortie of the two brothers (GNEISENAU) in the Atlantic Ocean. In this image, one can see the wear of the 51 gray after 2 months at sea. The previous camouflage pattern began to show through, the Baltic stripes visible on the hull midships. The recognition color painted atop the turrets was yellow RAL-1003. Both battleships put into the German occupied French port of Brest on 22 March 1941.

TIRPITZ entering the German occupied port of Gotenhafen, Poland, 14 March 1941. He was painted in the standard 50/51 pattern at that time, but differed from his brother BISMARCK in that the funnel cap was painted black.

GNEISENAU at sea near the end of Operation Berlin. Note the heavily worn paint on his hull, with portions of the previous camouflage pattern showing through, although the stern may have been painted slightly darker. The battleship was ridding high at the time of this photograph, possibly low on fuel and ammunition from their two month long sortie in the Atlantic Ocean.

Entering the harbor at the German occupied port of Brest, France, 22 March 1941, GNEISENAU looking well worn at the conclusion of Operation Berlin. The large, domed devices fitted midships were the 4m High Angle Directors for anti-aircraft fire control. These were triaxially stabilized on a gimbel mounting, so as the ship rolled and heaved to the wave action, the director was able to stay on a perfect horizontal plane at all times and in all directions.

This close-up of the forward superstructure and funnel of SCHARNHORST was taken at Brest a day, or two after both battleships arrived in late March 1941. Here is another view of the 4m HA AA directors, as well as the mounts they controlled, the twin 105mm AA guns situated on the 02 deck midships. Note the foretop main battery director with the addition installed on top, as well as the FuMO-22 radar antenna. The main and secondary gun turrets have canvas covers fitted to quickly conceal the recognition color painted atop those mounts. Note the midships AA director has the dome painted dark grey ?

BISMARCK in his Baltic camouflage pattern, while training with the heavy cruiser PRINZ EUGEN and briefly, TIRPITZ. Note the irregular painting of the stripes from port to starboard, when compared to the photograph below. This photo was taken in late April, early May 1941, near Gotenhafen.

In this photograph of BISMARCK, only the slopes to the Anton turret appear to be a lighter shade than the others ? This photo was taken from the aft deck of PRINZ EUGEN in early to mid-May 1941, with BISMARCK passing at high speed. A portion of the crew were mustered aft in this photo. Note that the rangefinder was removed and plated over on the Anton turret. Because these battleships were wet forward in anything above a moderate sea state, this device was prone to damage.

TIRPITZ, at the pier, in the German occupied port of Gotenhafen, Poland. He is nearly 100% finished and was working-up in the Baltic Sea at the date of this photograph, 5 May 1941. Adolf Hitler was visiting Germany's largest battleships, including BISMARCK, as well as her newest heavy cruiser PRINZ EUGEN, all on the same day. In this photograph, the crew of the TIRPITZ is mustered along the starboard main deck railing to receive their countries leader. Note the recognition color painted atop the main gun turrets was dark gray RAL-7016.

This image of the BISMARCK just as it arrived at Grimstadfjord, Norway, 21 May 1941. The crew had painted out the Baltic stripes on the superstructure en route from Gotenhafen, and are preparing to paint out the camouflage on the hull.

BISMARCK, photographed about mid-day, with the Baltic stripes painted over and the crew preparing to paint over the foreshortened bow and stern.

This poor quality photograph of the BISMARCK was taken as that battleship departed Norway and made its way out into the Atlantic Ocean for the first time, in company with the heavy cruiser PRINZ EUGEN, very late in the day, 21 May 1941.

All of the photographs on this page are of BISMARCK engaging the Royal Navy battlecruiser HMS HOOD and battleship PRINCE OF WALES in what was to become known as the Battle of Denmark Strait. BISMARCK and PRINZ EUGEN would sink HOOD and badly damage PRINCE OF WALES, 24 May 1941.

The damaged BISMARCK pulls alongside the heavy cruiser PRINZ EUGEN late on the morning of 24 May 1941. The battleship pulled ahead so that the cruiser could determine how much oil fuel was leaking from the battleship. At this time, the recognition paint color atop the main gun turrets was RAL-7016.

The hit in the bow of BISMARCK penetrated a fuel bunker and caused a severe leak. The battleship was down at the bow by 2° and had to reduce top speed to 27 knots. A decision was made to head for Brest, France for repairs, a journey that BISMARCK would never complete.

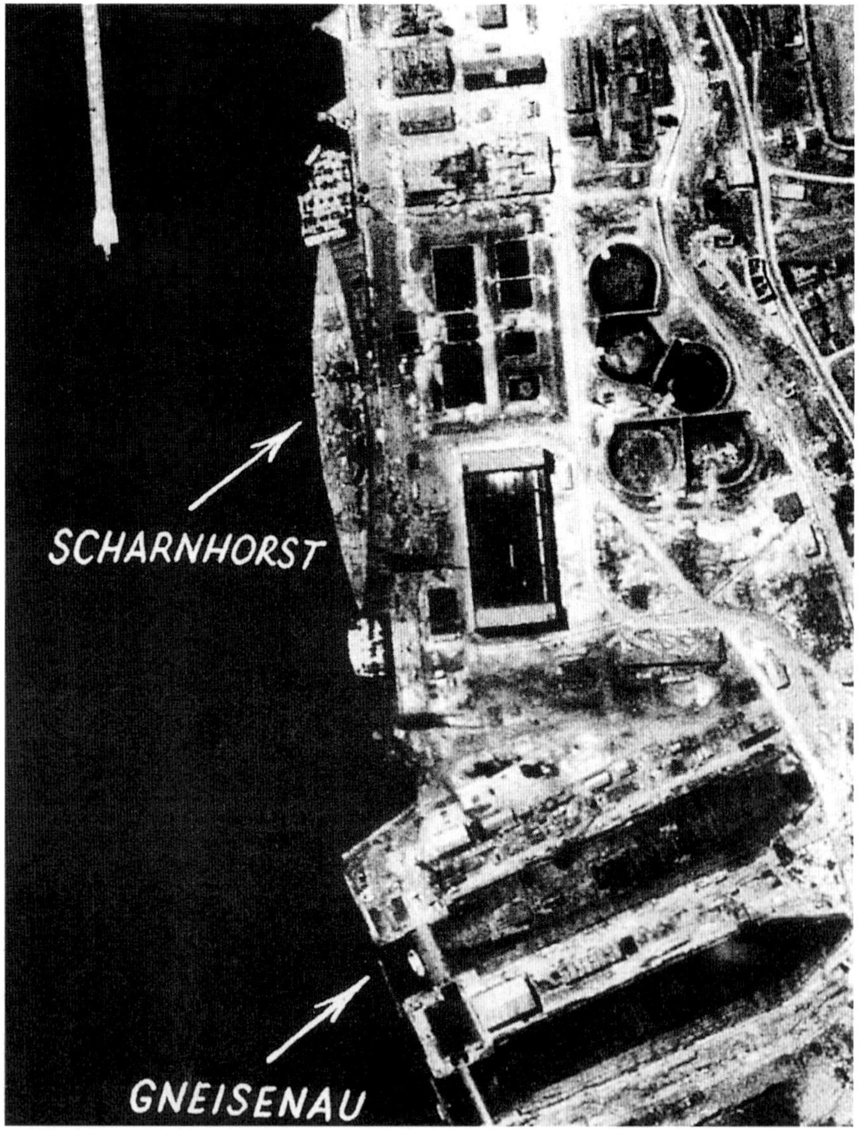

While at the German occupied port of Brest, France, SCHARNHORST and GNEISENAU were under constant surveillance and bombing from the British RAF. This action necessitated camouflaging the two battleships to the extent that they would not only be painted with a camouflage pattern, but also covered with camouflage netting. The photograph left and that at the top of the opposite page show the extent of the camouflage netting. GNEISENAU in the image above is almost completely hidden from view by the extensive use of camouflage netting.

TIRPITZ during his gunnery exercise in the Baltic Sea, June 1941. To this author's knowledge, the TIRPITZ was the only German battleship during the Second World War that did not carry a Baltic camouflage pattern. The battleship was painted in the standard 50/51 pattern with the tops of the main gun turrets painted dark gray RAL-7016. The main gun barrels and the funnel cap were painted black.

Another view of the TIRPITZ during gunnery training in the summer of 1941, in the Baltic Sea.

SCHARNHORST at the pier in Brest, France. Even at the pier, there was a lot of camouflage netting draped over the battleship in an attempt to conceal the warship from RAF bombers. Summer 1941.

This photograph was taken as SCHARNHORST was making a run down to La Pallice, from Brest, in an attempt to alleviate the amount of bombing he was under from the RAF. SCHARNHORST was hit by five bombs and later returned to Brest, as there was better AA protection there. One can see that some of the dark camouflage pattern remains in this image.

The left image is of GNEISENAU shortly after arriving at Brest, France. The battleship was still showing the wear from the two month long sortie and it would appear that the main gun turrets were covered with a dark canvas to conceal the recognition color painted on the top.

In the right image, GNEISENAU, in drydock, is mostly covered with camouflage netting and was also painted in a two-tone dark camouflage pattern consisting of RAL-7016 & 7024. The decks of the battleships were also painted at this time, possibly RAL-7024.

Close-up of the after superstructure and a portion of the funnel of the battleship TIRPITZ. This photograph was taken while the TIRPITZ was still working up and training in the Baltic Sea during the late summer of 1941. There were many problems discovered with his brother BISMARCK that were being addressed on TIRPITZ, which is why that battleship spent such a long period in the Baltic Sea, based out of Gotenhafen. In this photograph, three of TIRPITZ's twin 105mm AA mounts are visible, as well as one of his Arado Ar-196 floatplanes.

A view of the bridge and funnel of TIRPITZ, late summer 1941, while in the Baltic Sea. On the left side of this photo, atop the conning tower, as well as atop the bridge itself, are the main battery directors, fitted with the type FuMO-23 radar antenna. The German Navy used this as a fire control radar and not for surface, or air detection purposes.

Above is a view of the SCHARNHORST(L) and GNEISENAU(R) in drydock at Brest. They were kept in drydock to prevent torpedo attack and the drydock was drained so the battleships could not be sunk.

Right is an image of the bridge of GNEISENAU showing the application of camouflage painting and netting while at Brest.

TIRPITZ glides gracefully through the water as the battleship was then completed and ready for final crew training in the Baltic Sea. This photograph is dated sometime late in 1941. The paint colors used at that time was the standard 50/51 pattern.

Painted in a then new camouflage pattern consisting of dark grays, RAL-7024 on the majority of the hull and all of the superstructure, and RAL-7016 on the foreshortened bow and stern, TIRPITZ was made ready to depart German waters for Norway. This photograph was taken sometime in early January 1942.

GNEISENAU and SCHARNHORST in the harbor at the German occupied port of Brest, France, January, or February 1942. Note how "dirty" the two battleships were camouflaged. At this time the two were in preparation to depart Brest for Germany due to the non-stop bombing by the RAF. If they stayed, it would only be a matter of time until they would be destroyed, so the decision was made to relocate the battleships back to Germany. Note the small mast atop the after superstructure aboard GNEISENAU, also visible on page 40.

GNEISENAU in the process of re-fueling and replenishing ammunition for the return to Germany. A dark gray was used for the rough camouflage painting of both SCHARNHORST and GNEISENAU, possibly RAL-7024. Also, the turret tops were painted with Mittelblau 33.2, a pastel sky-blue color as a recognition marking.

The view from the bridge of the heavy cruiser PRINZ EUGEN during Operation Cerberus, also known as the "*Channel Dash*" by the British. Ahead, GNEISENAU, and leading the squadron was SCHARNHORST. The warships and their escort of destroyers and torpedo boats departed Brest at 2300hrs., 11 February 1942. As can be seen in this photo, the squadron was moving at high speed in a bold daylight dash up the English Channel, a move un-anticipated by the British.

12 February 1942, SCHARNHORST at high speed in the English Channel during Operation Cerberus, a sortie of the two battleships and the heavy cruiser PRINZ EUGEN to evacuate those capital ships from France back to Germany.

GNEISENAU, also on 12 February, during Operation Cerberus. The sortie was not noticed by the British until the German squadron had reached the Strait of Dover, nearly the half-way point of the operation, catching them quite by surprise, in the middle of the day! Note the then new aircraft hangar recently constructed while at Brest. He was also to have the mainmast relocated to the after superstructure as in SCHARNHORST, which would have made the two battleships very similar to each other.

GNEISENAU, photographed from PRINZ EUGEN during the "*Channel Dash*." He hit a mine just after passing Amsterdam, at 2056hrs., although slowed, the battleship was able to continue at speed onwards to Germany.

SCHARNHORST hit a mine at 1530hrs., three hours after passing Dover. This caused some boilers to go off-line, but the battleship was able to continue and soon built up speed to rejoin the squadron. Again, he struck another mine, this time at 2235hrs, on the approach to German waters. SCHARNHORST went dead in the water for 45 minutes while the crew scrambled to get the battleship moving again. He steamed into Wilhelmshaven at 14 kts., at 1230hrs. on 13 February 1942, remarkable considering the damage.

SCHARNHORST viewed from astern as he proceeded to Wilhelmshaven. Although the operation was somewhat successful, the move was an evacuation of military forces, signalling a retreat, or setback for Germany. The British by their continual aerial bombing of Brest, and the lack of support from Germany's air force, the Luftwaffe, scored a major victory here. They had deprived Germany of a port with better access to the Atlantic Ocean and their future commerce raiding operations.

27 February 1942, just 14 days after the conclusion of Operation Cerberus, GNEISENAU in drydock, undergoing an inspection for repairs from the mine damage in the English Channel, was hit in the bow by a large bomb in a raid by the RAF. Against standard procedures, the ammunition was not removed when the battleship was docked. The bomb hit set off a chain reaction, causing a flame-out of the ready ammunition in and below turret Anton, causing major damage and completely disabling the battleship. 112 crew were killed, and as one can see by this aerial photograph, the bomb hit left a substantial hole in the bow of the GNEISENAU.

TIRPITZ in Norwegian waters, 6 March 1942, departing for the North Atlantic to attack the Allied convoys known to be in that area. The battleship was painted in a very dark camouflage pattern of RAL-7024 and the foreshortened bow and stern were RAL-7016, with funnel cap painted black. The decks appear to have been painted RAL-7024.

Another view of TIRPITZ, at this time he was moored in Fättenfjord, Norway, up against the mountainous shore in an effort to hide the battleship. The turret tops were covered with canvas to hide the recognition paint color.

An aerial view of the TIRPITZ underway in one of the fjords in Norway. One can see that the decks were painted dark, possibly RAL-7024.

The primary photograph on this page is of the bridge structure of the TIRPITZ just after Operation Sportpalast against Allied convoys in the North Atlantic Ocean. The battleship was anchored at Bogen, Norway for a few days during mid-March 1942, later returning to Fättenfjord, near Trondheim.

TIRPITZ was back at Fättenfjord, this photo taken in the early summer of 1942. Note the use of canvas and netting in the beginning stages of trying to camouflage the battleship from aerial observation. TIRPITZ appears to retain his last camouflage paint pattern at this time.

This view of the battleship TIRPITZ was taken on 3 June 1942, just as efforts to apply a new camouflage paint pattern were completed. It was decided that the camouflage of TIRPITZ should try to hide the battleship against the mountainous background he was moored near. Green and browns were used, possibly Dunkelgrün 35.3, Hell Braun 36.1 and Dunkel Braun 36.3. The deck was also camouflaged using a selection of the same colors.

With this then new camouflage pattern on the TIRPITZ, the bow and stern were painted with Schnellbootweiss, an off-white primarily used on torpedo boats. Both this and the photograph above were taken in late June 1942 at Fättenfjord, Norway.

By July 1942, TIRPITZ was re-painted with a new type of camouflage pattern, known as "Disruptive." The port side of the battleship was painted in a "Splinter" and the starboard side in a "Dazzle" pattern. The paint colors used were RAL-7001(50), 7024, 7016 and Schnellbootweiss.

TIRPITZ, center, with the heavy cruiser ADMIRAL HIPPER and three destroyers assembled just prior to Operation Rösselsprung in Altafjord, early July 1942.

TIRPITZ leading ADMIRAL HIPPER, ADMIRAL SCHEER, 12 destroyers and torpedo boats and a fleet oiler, out to the North Atlantic Ocean for Operation Rösselsprung, 5 July 1942. This was an operation against the Allied supply convoy PQ-17 bound for Murmansk, Russia.

The German squadron returned to Altafjord with Operation Rösselsprung terminated due to highly successful U-boat operation against convoy PQ-17. Both the primary photograph and the smaller image on this page were taken from the deck of the heavy cruiser ADMIRAL HIPPER looking at the squadron, including the battleship TIRPITZ, arriving back to Norway. The image right was taken while ADMIRAL HIPPER was re-fueling from TIRPITZ. The German Navy performed underway refueling in-line, with the supply vessel leading.

Both photographs on this page are of the TIRPITZ starboard side and his dazzle pattern camouflage. The paint colors used were the same as those on the port side, RAL-7000(51), 7016, 7024 and Schnellbootweiss. The wood decks were natural, which tended to weather dark and towards gray and the steel decks were dark gray.

TIRPITZ sometime after Operation Rösselsprung, July through September 1942, showing the starboard splinter camouflage, anchored in the Bogenfjord, Norway.

Above, TIRPITZ at anchor in Bogenfjord, summer 1942. This photograph was taken by RAF Aerial Reconnaissance.

Below is an image of TIRPITZ on a rare visit to Narvik, late summer 1942.

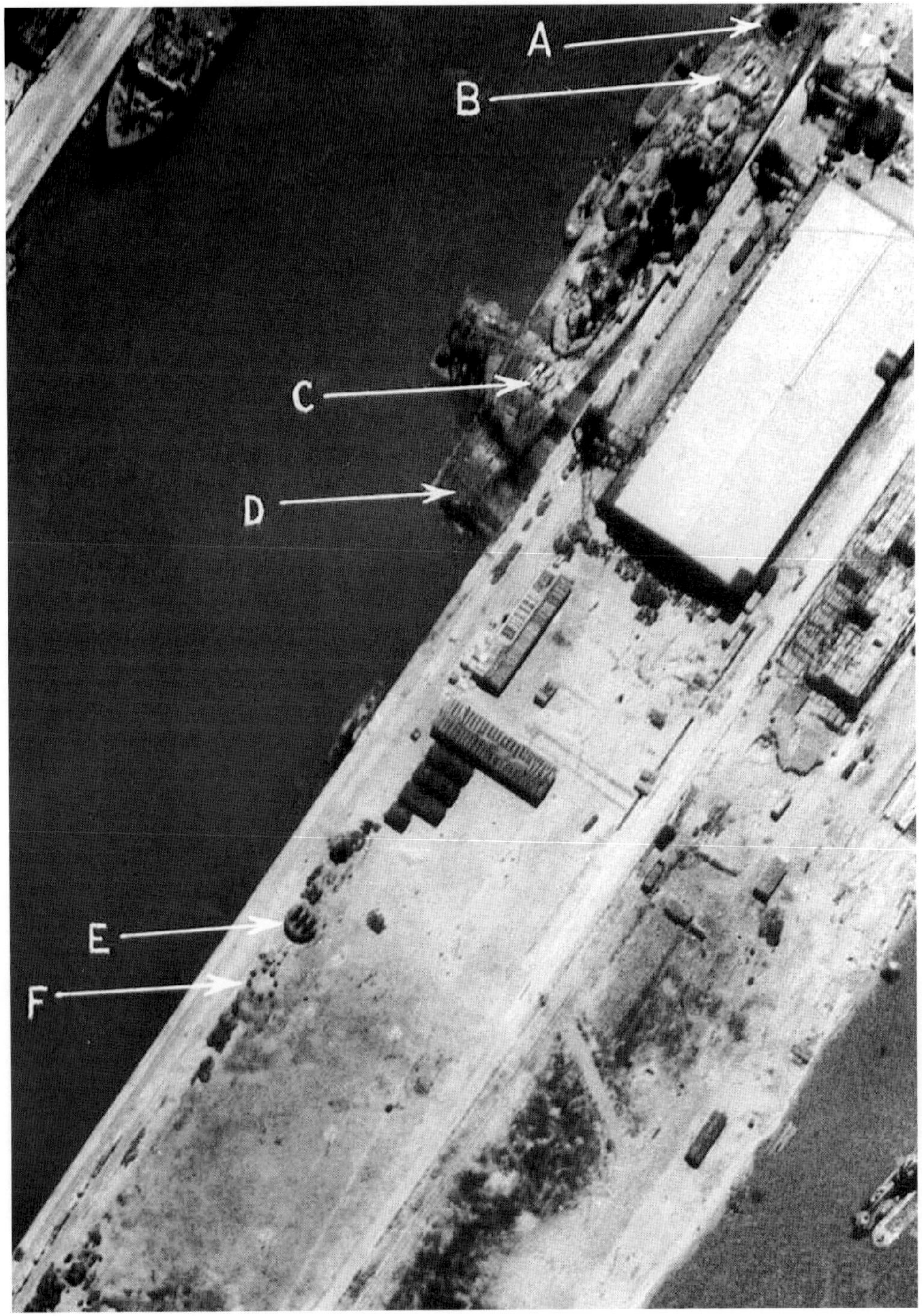

GNEISENAU, the forgotten battleship, still in existence, at the German occupied port of Gotenhafen, Poland. The photograph to the left shows the battleship in a state of disassembly for the planned re-construction, an RAF aerial reconnaissance photo from July 1942. The photograph below was another RAF aerial reconnaissance image taken in October 1942 showing GNEISENAU in a floating drydock in the process of having the bow removed. Note how the battleship was still able to raise steam.

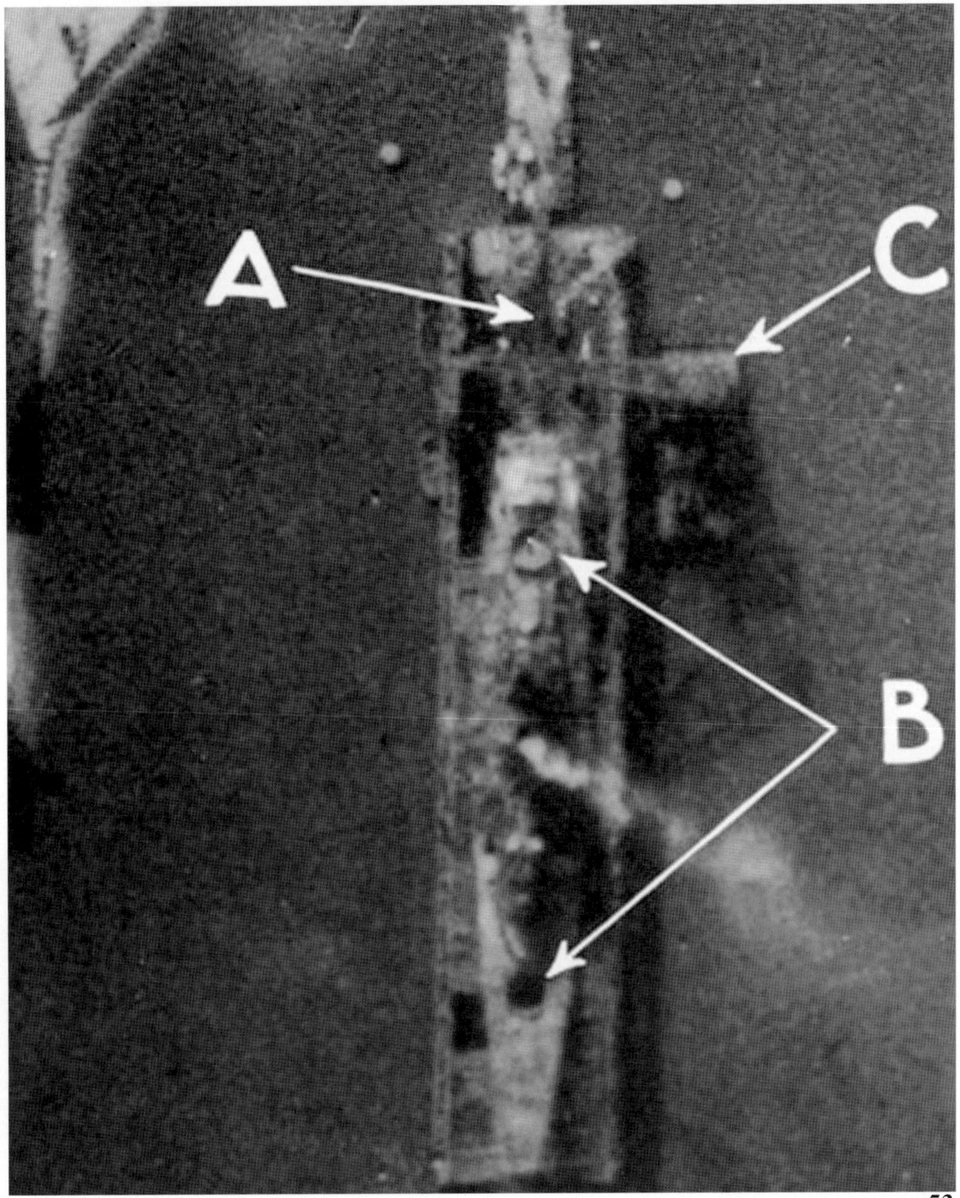

Tied up to a pier at Gotenhafen, Poland, behind two rows of anti-torpedo nets, the SCHARNHORST was in the finishing stages of repair and refit after the damages inflicted during Operation Cerberus. The date of this photograph is believed to have been sometime during October 1942. The battleship received upgrades and additions to his radar systems and anti-aircraft gun complement.

SCHARNHORST photographed sometime in November 1942 while on post-refit trials and crew training in the Baltic Sea, based out of Gotenhafen. The then new camouflage pattern consisted of RAL-7024 overall, and the foreshortened bow and stern painted with the Schnellbootweiss.

This photograph of the SCHARNHORST was taken sometime late in 1942 at Gotenhafen, while that battleship was operating in the Baltic Sea. The damage that SCHARNHORST received from hitting the two mines in the "*Channel Dash*" was extensive and was not fully repaired. Repairs were performed to temporally fix the problem, but he needed a lot more yard time. This was never to be completed. During this time, the battleship did receive an additional FuMo-26 fire control radar, with the antenna fitted to a modified foretop, which was then square in shape. The previous foretop modification was cylindrical in shape.

SCHARNHORST en route to Norway, 6 through 9 March 1943. This was the third, and successful attempt for SCHARNHORST to transit to Norway, the first two failing due to air attack and problems refueling the destroyer escort.

This was an image of the SCHARNHORST as that battleship arrived at Bogen, near Narvik, Norway. By the bow and stern wakes, one can see that the battleship was moving at high speed. SCHARNHORST made this voyage with three destroyers as escort.

SCHARNHORST and the TIRPITZ together in the Ofotfjord, March 1943. SCHARNHORST's camouflage pattern was deceptive, while that on TIRPITZ was disruptive.

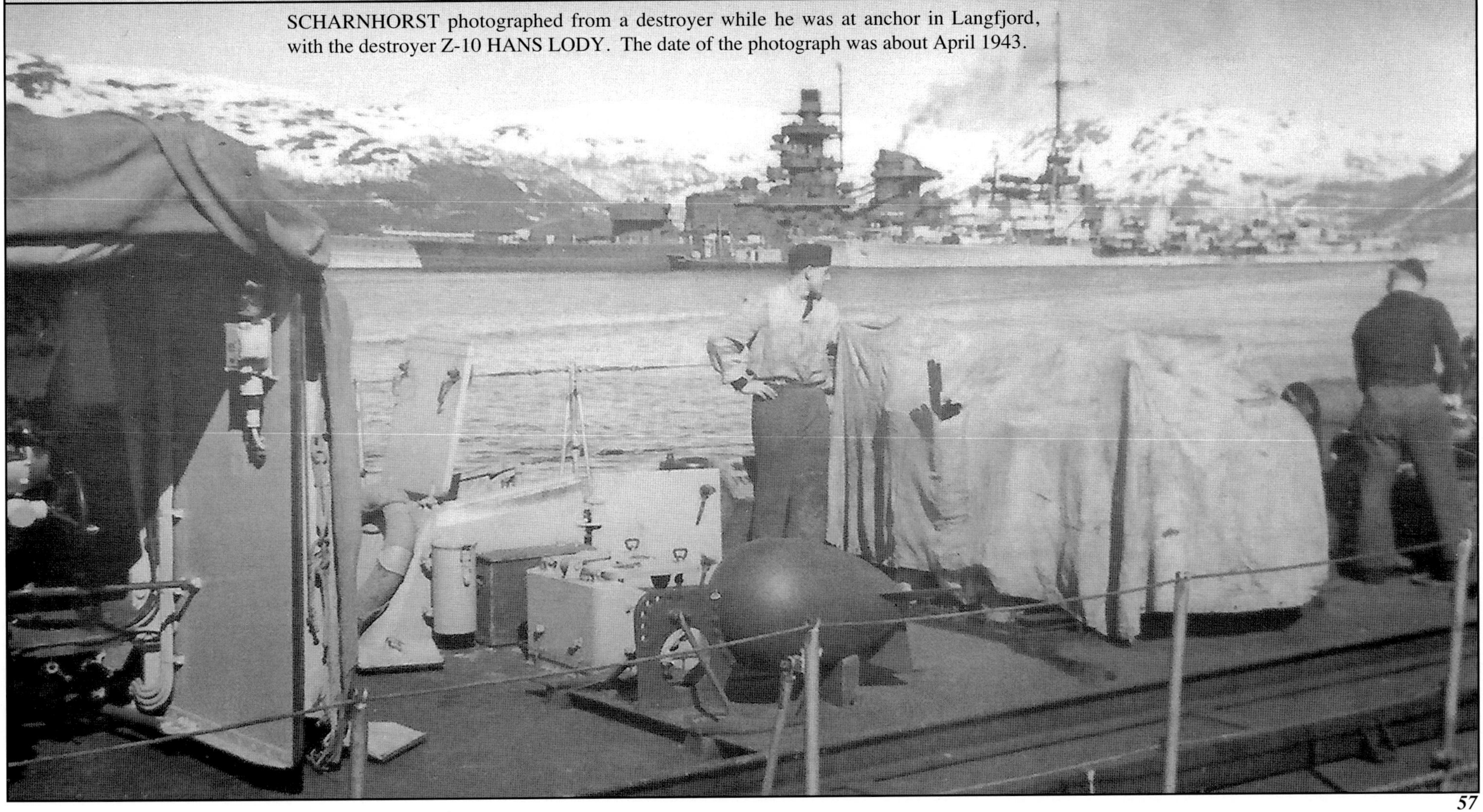

SCHARNHORST photographed from a destroyer while he was at anchor in Langfjord, with the destroyer Z-10 HANS LODY. The date of the photograph was about April 1943.

In this smaller photo of TIRPITZ, the three shades of gray used for camouflage can be seen. The crew was mustered as the SCHARNHORST passed alongside.

The primary photo on this page is of the SCHARNHORST anchored at Ofotfjord, near Narvik, Norway, March 1943. The photograph was taken from the TIRPITZ as SCHARNHORST past alongside. The poor quality of the image is due to humidity having damaged the original film.

TIRPITZ, SCHARNHORST and LÜTZOW at Bogen at the end of March 1943. Compare to the image below. In this photo, TIRPITZ wore the dazzle pattern on his starboard side and in the photo below, he wore the splinter pattern on his port side.

TIRPITZ anchored in the Kaafjord, near Alta, April 1943. On the far left of this image is the heavy cruiser LÜTZOW at anchor. Kaafjord was just out of range of RAF land based bombers, and closer to the Allied convoy routs to Murmansk, Russia, but also far enough North to be a problem during the long winters. TIRPITZ wore a dark gray RAL-7016 recognition paint color atop the turrets at this time.

Battleship SCHARNHORST and the destroyer Z-15 in front of him, Langfjord, April 1943. The camouflage of Z-15 was designed to confuse, and here, one can see that it worked.

The inset photo is of GNEISENAU at Gotenhafen, mid-1943, sitting inactive.

TIRPITZ, known as the "Lonely Queen of the North," in a fjord in northern Norway, June 1943. The nick-name was of British origin, as Germans referred to their warships as masculine.

Both this top image and the small inset image of the battleship SCHARNHORST were photographed during July 1943. During the summer of 1943, the battleships trained together in Norwegian waters.

TIRPITZ slowly returning to his anchorage in the Kaafjord during August 1943. During July, this battleship received a new camouflage pattern, using the same paint colors as with the previous patterns, but was painted with a new splinter pattern, and this time on both port and starboard sides.

TIRPITZ, and just beyond, left, SCHARNHORST en route to the island of Spitsbergen, within the Arctic Circle, to bombard the Allied weather station there in Operation Sizilien.

TIRPITZ led this operation in company with the battleship SCHARNHORST and 9 destroyers. They bombarded the Allied weather station and landed a ground force to ensure total destruction of this Allied base.

SCHARNHORST and TIRPITZ retuning to Kaafjord, Norway, 9 September 1943. TIRPITZ would remain in the Kaafjord, while SCHARNHORST would later move to the Langfjord. Later that month, TIRPITZ would be successfully attacked by British midget submarines, causing severe damage.

This photograph of the SCHARNHORST shows him in the then new camouflage pattern that would be his final paint scheme. It was very different from that on TIRPITZ, left and below, but used the same paint colors.

Below is an image of the battleship TIRPITZ during the repair period from the damage inflicted by the Royal Navy X-Craft midget submarine attack. One can see that the splinter camouflage pattern and colors worked well to conceal and distort the shape of the big battleship.

This starboard side view of SCHARNHORST shows the camouflage pattern in the area of the hull, especially the stern and bow areas. It has been determined from other photography available, that the pattern was symmetrical from port to starboard.

The paint colors used on the SCHARNHORST were more than likely grays. Those paints would have been RAL-7000, 7024, 7016, 9005 and Schnellbootweiss. Much conjecture has been used as to the paints utilized for this pattern, from browns/greens/grays and grays/blues to just the gray tones. Due to supply difficulties and past patterns utilized, the gray pattern makes the most sense. As of the date of this publication, no document, or color photography has been discovered to tell us the exact paints used.

Above, is one of the last known photographs of the battleship SCHARNHORST, as it returned from Operation Sizilien, bombarding the Allied weather station on the island of Spitsbergen. SCHARNHORST would depart Norway for the last time on 25 December 1943, in Operation Ostfront to attack the Allied convoy to Murmansk, Russia. He was ambushed and sunk by the Royal Navy north of the Arctic Circle.

Below, the TIRPITZ on a post-damage repair trial in Altafjord, Norway, mid, to late March 1944. During this repair period, September 1943, through March 1944, TIRPITZ also had most of his radar systems up-graded, as well as additional new radar systems installed. From the photograph below, one can see that TIRPITZ had a new camouflage pattern painted on, as it had been some time before the repair work was completed.

Above and below, the TIRPITZ on a post-damage repair trial in Altafjord, Norway, mid, to late March 1944.

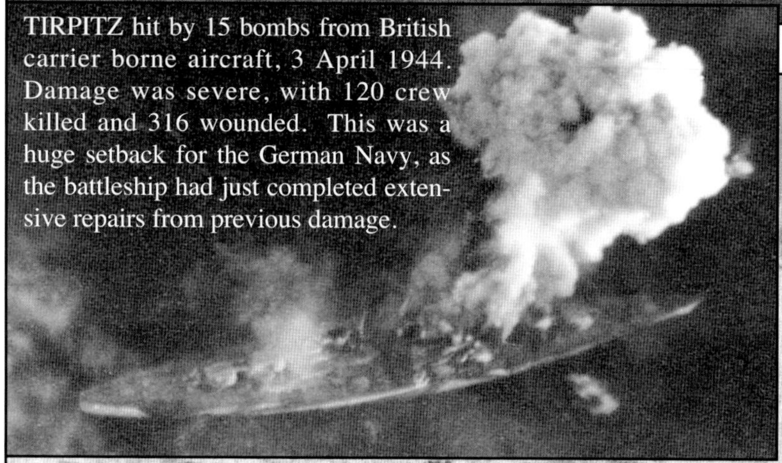

TIRPITZ hit by 15 bombs from British carrier borne aircraft, 3 April 1944. Damage was severe, with 120 crew killed and 316 wounded. This was a huge setback for the German Navy, as the battleship had just completed extensive repairs from previous damage.

The bow and stern of TIRPITZ were painted with Schnellbootweiss in this camouflage pattern.

An aerial view of the TIRPITZ, taken sometime after the aircraft carrier borne attack. He was re-painted in another camouflage pattern at that time. This pattern was extended to the main deck, consisting of RAL-7024 on the natural wood decking. Note the massive amount of fuel oil leaking from the damaged battleship.

This view of the TIRPITZ was taken while that battleship was undergoing post-damage repair trials in Altafjord during early July 1944. The paint colors used in the camouflage seen in this photograph were RAL-7016 on the hull, first deck level of the superstructure, and main gun turrets Anton and Dora. From that horizontal height, upwards, to the base of the Foretop Control Stand platform, the battleship was painted with RAL-7000(51). From the Control Stand platform, upwards, the remaining foretop was painted Schnellbootweiss. Due to the damage, the starboard boat and aircraft crane was removed, seen in this photograph, and above. Because of that, the ship's boats could no longer be carried on the starboard side of the battleship.

This close-up of an aerial view of the naval installation at the German occupied port of Gotenhafen, Poland, one of two of Germany's primary naval bases in the Baltic Sea, shows the out of commission battleship GNEISENAU. He was in storage at this time, late 1944, with all work suspended. Note the absence of any bow and main gun turrets. The German battleship was to undergo an extensive modification and repair, but this was never carried out. The light cruiser tied up alongside was the LEIPZIG. This photograph was taken by a British RAF reconnaissance aircraft.

TIRPITZ at anchor performing a boiler test, in the Altafjord, Norway, late July 1944, just prior to his last venture out to the North Atlantic with 5 destroyers as escort. Note the upgrade to the FuMO-26 Main Battery Fire Control Radar antenna on the after rangefinder.

The end of the TIRPITZ, 12 November 1944. A total of 31 RAF Lancaster bombers dropped 30 "Tallboy" bombs on the German battleship, scoring three hits and several near misses. The "Tallboy" weighed 12,000 lbs. and caused catastrophic damage to the battleship, causing a magazine explosion and the TIRPITZ to capsize in shallow water. About 1,000 of the 1,900 men aboard were killed, with 87 crew rescued by cutting through the upturned bottom to release them from the capsized battleship, in a very similar fashion as performed on the American battleship OKLAHOMA at the bombing of Pearl Harbor in December 1941. This photograph was taken from a far shore in the Sorbotn Fjord, near Tromso, Norway, just before the TIRPITZ exploded and capsized.

The upturned hull of the last operational German battleship sunk, the TIRPITZ, taken from a British RAF reconnaissance aircraft, just days after the fatal attack.

The scuttled wreck of the battleship GNEISENAU. The de-commissioned battleship was towed out to the harbor entrance and scuttled as a block-ship to deny the Russian Navy entrance to the harbor, 27 March 1945. With his scuttling, GNEISENAU was the last of the four battleships of the German Navy to be sunk.

Sometime prior the scuttling of the last German battleship GNEISENAU, all of the armament, rangefinders and directors were removed. One can see in this photograph and the one on the previous page the large hangar built at Brest, France. Some work on his conversion was performed, as the mainmast was removed from the backside of the funnel. It was to have been relocated as that on SCHARNHORST, to the after superstructure. So with the scuttling of the GNEISENAU, the first battleship completed, ended the battleship force of the German Navy.

REFERENCES

Battleship Bismarck
U. Elfrath & B. Herzog, Schiffer Publishing LTD., 1989

Battleships of the Bismarck Class
G. Koop & K. P. Schmolke, Greenhill Books, 1998

Battleship Gneisenau
S. Breyer, Schiffer Publishing LTD., 1990

Battleship Scharnhorst
R. Shoker, Oxford Museum Press, 1999

Battleships of the Scharnhorst Class
G. Koop & K. P. Schmolke, US Naval Institute Press, 1999

Battleship Tirpitz
S. Breyer, Schiffer Publishing LTD., 1989

Chronology of the War at Sea 1939-1945
J. Rohwer, US Naval Institute Press, 2005

German Naval Camouflage - Vol. 1 1939-1941
J. Asmussen & Eric Leon, US Naval Institute Press, 2012

The Bismarck
P. Caresse, Lela Press, 2004

The Scharnhorst
P. Caresse, Lela Press, 2005

Tirpitz - Hitler's Last Battleship
J. Asmussen & K. Akra, Midt-Troms Museum, 2006

RESOURCES

Snyder & Short Enterprises
PMB224 - 9175 Kiefer Blvd., Sacramento, CA 95826-5147
Web Site: www.shipcamouflage.com

US Naval Historical Center
805 Kidder Breese SE, Bldg. #57
Washington Naval Yard, Washington DC, 20374-5060
Still Photos (202)433-2765•Web Site: www.history.navy.mil

US National Archives
8601 Adelphi Rd., College Park, MD 20740-6001
Web Site: www.nara.gov

Internet Sources:
www.bismarck-class.dk
www.kbismarck.org
http://en.wikipedia.org/wiki/kriegsmarine
www.forum-marinearchiv.de
www.modelwarships.com
www.NavWeaps.com
www.pixpast.com
www.SteelNavy.com
www.warshipmodelsunderway.com

USS INDEPENDENCE CVL-22, the subject of the next book in the Warship Pictorial Series.

ACKNOWLEDGMENTS

CLASSIC WARSHIPS
would like to express it's gratitude for assistance from the following individuals -
John Asmussen • Antonio Bonomi • David Saxton
Thomas Schmid • Randy Shoker • Siegfried Venzke

WARSHIP PICTORIAL SERIES

WP 4 - USS TEXAS BB-35
WP 10 - INDIANAPOLIS & PORTLAND
WP 20 - HMS HOOD
WP 23 - Italian Heavy Cruisers
WP 24 - ARLEIGH BURKE Class Destroyers
WP 30 - IJN TAKAO Class Cruisers
WP 31 - USS BUCHANAN DD-484
WP 32 - SOUTH DAKOTA Class Battleships
WP 33 - USS LEXINGTON CV-2
WP 34 - USN Battleships in Color
WP 35 - TICONDEROGA Class Cruisers
WP 36 - Kriegsmarine SCHARNHORST
WP 37 - RM ROMA
WP 38 - IJN NAGATO Class Battleships
WP 39 - German Battleships 1939-1945

AIRCRAFT PICTORIAL SERIES

AP 1 - MIDWAY Air Wings
AP 2 - SB2U Vindicator
AP 3 - OS2U Kingfisher
AP 4 - F4F Wildcat

FRONT COVER: KM GNEISENAU, in a collage created from a early 1939 black and white photograph, with portions of SCHARNHORST added and then colorized to represent GNEISENAU as he was in late 1939.

BACK COVER: KM TIRPITZ, at speed in the Altafjord, Norway, March 1944, in a colorized black and white photograph.